# Wireless UWB Body Area Networks

# Wireless UWB Body Area Networks

## Using the IEEE802.15.4-2011

Matti Hämäläinen

Jari Iinatti

Editor

Lorenzo Mucchi

Academic Press Library in Biomedical Applications of Mobile and Wireless Communications

ELSEVIER

AMSTERDAM • BOSTON • HEIDELBERG • LONDON
NEW YORK • OXFORD • PARIS • SAN DIEGO
SAN FRANCISCO • SINGAPORE • SYDNEY • TOKYO
Academic Press is an imprint of Elsevier

AP

Academic Press is an imprint of Elsevier
32 Jamestown Road, London NW1 7BY, UK
The Boulevard, Langford Lane, Kidlington, Oxford, OX5 1GB, UK
Radarweg 29, PO Box 211, 1000 AE Amsterdam, The Netherlands
225 Wyman Street, Waltham, MA 02451, USA
525 B Street, Suite 1900, San Diego, CA 92101-4495, USA

First published 2014



**British Library Cataloguing-in-Publication Data**
A catalogue record for this book is available from the British Library

**Library of Congress Cataloging-in-Publication Data**
A catalog record for this book is available from the Library of Congress

ISBN: 978-0-12-800931-4



This book has been manufactured using Print On Demand technology. Each copy is produced to order and is limited to black ink. The online version of this book will show color figures where appropriate.

# CONTENTS

# LIST OF ABBREVIATIONS AND VARIABLES

**Acronyms**

| | |
|---|---|
| A-rake | all rake receiver |
| ARQ | automatic repeat request |
| ASK | amplitude shift keying |
| AWGN | additive white Gaussian noise |
| BER | bit error rate |
| BPM | burst position modulation |
| BPSK | binary phase shift keying |
| CC | convolutional code |
| CSMA-CA | carrier-sense multiple access-collision avoidance |
| CSS | chirp spread spectrum |
| DSSS | direct sequence spread spectrum |
| E2C | environment-to-cloud |
| ECC | error control coding |
| ED | energy detector |
| FEC | forward error control |
| FFD | full function device |
| GFSK | Gaussian frequency shift keying |
| H2M | human-to-machine |
| IEEE | the Institute of Electrical and Electronics Engineers |
| IFFT | inverse fast Fourier transform |
| IoT | Internet of Things |
| IR-UWB | impulse radio based ultra wideband |
| ISM | industrial, scientific, and medical |
| LECIM | Low Energy Critical Infrastructure Monitoring |
| LOS | line-of-sight |
| M2H | machine-to-human |
| M2I | machine-to-infrastructure |
| M2M | machine-to-machine |
| MAC | medium access control |
| MPSK | M-ary phase shift keying |
| MSK | minimum shift keying |
| NIST | National Institute of Standards and Technology |
| OOK | on–off keying |

| | |
|---|---|
| O-QPSK | offset quadrature phase shift keying |
| PAN | personal area network |
| PER | packet error rate |
| PHY | physical layer |
| P-rake | partial rake receiver |
| RFD | reduced function device |
| RFID | radio frequency identification |
| RMS | root mean square |
| RS | Reed–Solomon code |
| SAR | specific absorption rate |
| SNR | signal-to-noise ratio |
| S-rake | selective rake receiver |
| TDMA | time division multiple access |
| UWB | ultra wideband |
| WBAN | wireless body area network |
| WBSN | wireless body sensor network |
| WLAN | wireless local area network |
| WPAN | wireless personal area network |
| WSN | wireless sensor network |

**Variables**

| | |
|---|---|
| $b(t)$ | locally generated reference burst |
| $E_b$ | signal energy |
| $f_c$ | center frequency |
| $g_0^{(k)}$ | position modulated bit |
| $g_1^{(k)}$ | phase modulated bit |
| $h(t)$ | channel impulse response |
| $k$ | transmission interval |
| $l^{(k)}$ | burst hopping position |
| $M$ | number of distinguishable multipath components |
| $N$ | number of slots in a symbol quarter |
| $N_{cpb}$ | number of pulses per burst |
| $N_0$ | zero mean Gaussian noise |
| $n$ | index |
| $n(t)$ | white Gaussian noise |
| $p(t)$ | transmitted pulse waveform |
| Q1 | first quarter |
| Q3 | third quarter |
| $R_d$ | data rate |
| $R_s$ | symbol rate |

| | |
|---|---|
| $r(t)$ | received signal (burst) |
| $s$ | transmission slot |
| $T_{BPM}$ | half of the symbol length |
| $T_{burst}$ | burst length |
| $T_c$ | chip length |
| $T_{opt}$ | optimal additional integration time |
| $T_s$ | symbol length |
| $t$ | time |
| $v_0$ | decision variable for bit "0" |
| $v_1$ | decision variable for bit "0" |
| $v_i$ | decision variable |
| $w_i$ | decision variable |
| $x(t)$ | transmitted signal (burst) |

CHAPTER 1

# Introduction



Measuring human physiological signs remotely requires the sensor nodes attached to the human body or to clothes, and a reliable access to a network outside of a human body. To satisfy this need, several communication protocols and standards have been developed and defined during the previous years. Wireless sensor network, wireless body sensor network, or wireless body area network (WSN, WBSN, and WBAN, respectively), as you wish to say, have been seen as a future way to carry out a remote monitoring function of humans' vital signs. As the variation in naming indicates, the interest toward the local networks utilized close to human body communication has been seen as an interesting approach by several authors and study groups. WSN consists of sensor nodes, which have detector(s), a radio module, and a power source. The special case of WSN is WBAN (or WBSN in other name), which is used to create a network around a human body

Wireless UWB Body Area Networks. DOI: http://dx.doi.org/10.1016/B978-0-12-800931-4.00001-6

and set up wireless connections outside of the body, on a body surface, or even inside a body (Figure 1). In addition, it is possible to implement actuators into these body sensor nodes if needed. Those actuators in medical WBANs can be such as insulin pumps or other medicine delivery devices. In WBAN, all data transmission within the sensor and body network is carried out wirelessly, which means not only easy installation of the network components but also more comfortable and light carry on devices. It is also possible for some sensors to connect body hub, i.e., the local access point in a body, via cables. The operation range of WBAN applications is not exceeding few meters. By extension of humans only, WBAN type approaches can be utilized with animals, in machine-to-machine (M2M), machine-to-infrastructure (M2I), human-to-machine (H2M), etc. remote monitoring and controlling applications.

In general, the applicability of WSN is extremely wide. The next step is going to the "Internet of Things" (IoT) when huge amounts of electrical appliances are connected together and thus producing an extremely lot of data. Also monitoring environmental changes can lead to a environment-to-cloud (E2C) solution where monitoring results are put to cloud for further considerations. The use of cloud services allows ubiquitous data access for all the authorized persons independently on their location. This opens new ways to monitor patients, elderly, etc. remotely and thus, providing them a safe way to stay out

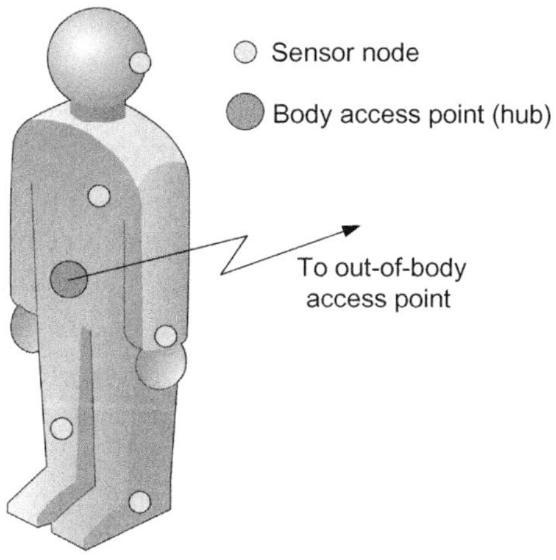


*Figure 1 Example of body area network components.*

of any care institute, which increases their quality of life and finally decreases the overall healthcare costs.

For short-range communications, there are many wireless technologies and standards available that can be used in numerous applications. The most common solutions are WLAN IEEE 802.11-family [1], Bluetooth IEEE 802.15.1 [2], ZigBee IEEE 802.15.4 [3,4], and its amendment, which is utilizing ultra wideband (UWB) as a physical layer (PHY) technology—IEEE 802.15.4a [5]. The first three technologies are using the crowded industrial, scientific, and medical (ISM) band, which calls for thoroughly to take into account interference and coexistence issues due to the heavy usage of that operative frequency band. On the other hand, a UWB signal is spread over a very large frequency range, more than 500 MHz with very low power spectral density, and thus it is an underlay approach for other existing radio systems. From a sensor networking point of view, both the wireless local area network (WLAN) and Bluetooth tend to have rather high power consumption, which leads to restrictions to battery life of a sensor node. ZigBee, and especially UWB technology, will fight against this problem by introducing devices with lower power consumption. Moreover, the newest version of Bluetooth, aka Bluetooth Low Energy (Bluetooth v4.1), is decreasing the power consumption of Bluetooth devices still retaining the backward compatibility with older versions of the Bluetooth standard. For medical, welfare, consumer, etc. WBAN applications, there is also dedicated standard IEEE802.15.6-2012 [6] available. The trend is going toward ultralow power appliances, requiring energy optimized protocol stacks and low power radio combined with small size and weight.

This book discusses the usability of UWB technology defined by the IEEE802.15.4-2011 [7] (former known as IEEE802.15.4a) in the WBANs used in healthcare and welfare related applications from a communication point of view. Other applications than communication, such as ranging and location tracking, are not covered here. The book first gives a short overview on the IEEE802.15.4-family and then touches IEEE802.15.4-2011 based UWB solution in more detail. According to the results discussed, IEEE802.15.4-2011 based WBAN can be used as a communication network in this kind of solution targeted for personal healthcare applications. The technology behind this standard has matured enough so it is available on the shelf market, which makes it a potential solution for the dedicated commercial WBAN systems. As a comparison to the newer IEEE802.15.6-2012 standard [6] for WBAN, those markets are not opened yet and the products based on the new standard are on the way to the market.

Remote and self-monitoring are the approaches how the future health care services are going to be organized. By giving people this possibility, it is conceivable to reallocate nursing staff to more urgent tasks. The demographics forecast is predicting that the ratio between the population who are actively in working life and those who are retired or are otherwise out of employment is getting smaller, and thus, new ways to give medical support with less people need to be developed. WBANs are one implementation in this process to make healthcare processes more effective and self-driven. This chapter is devoted to IEEE802.15.4-2011 (and especially the previously known IEEE802.15.4a part of it) based WBAN systems. The novel IEEE802.15.6-2012 [6] based WBAN approaches are not covered in this book.

## 1.1 IEEE802.15.4 STANDARD FAMILY

In this section, the IEEE802.15.4 standard family is briefly introduced. The standard family is targeted for low data rate applications, which can be used in short-range wireless personal area networks (WPANs) as well as WBANs. The first applications utilizing the IEEE802.15.4 standard were M2M and M2H communications, but the utilization areas for the standard are much wider. The standard family includes in the original IEEE802.15.4 standard from 2003 [3] but also its evolving amendments. Within the standard document and its amendments, PHY and medium access control (MAC) layer implementations and interconnections were carefully described. The standard was evolving in such a way that starting from 2.4 GHz operation, the first amendment (IEEE 802.15.4-2006 or IEEE 802.15.4b) version made available operations in 868 and 915 MHz bands [4]. As the original IEEE Std 802.15.4™-2003 was based on time division multiplexing and binary phase shift keying (BPSK) or offset quadrature phase shift keying (O-QPSK) modulation, the IEEE Std 802.15.4™-2006 introduced direct sequence spread spectrum (DSSS) with O-QPSK, BPSK, and amplitude shift keying (ASK) modulations. In the next phase, a new PHY technology, impulse radio based ultra wideband (IR-UWB), was taken into account and an amendment for an alternate PHY layer was accepted in 2007 [5]. In addition to IR-UWB, the IEEE802.15.4a was supporting chirp spread spectrum (CSS) PHY.

The amendments IEEE802.15.4c [8] and IEEE802.15.4d [9] were targeted for frequency range 779–787 MHz used in China and 950 MHz band used in Japan, respectively. In addition to the new frequency bands, these amendments introduced M-ary phase shift keying (MPSK) for Chinese and Gaussian frequency shift keying (GFSK) for Japanese markets. Both standards saw the light in 2009.

All the amendments before 2011 have been collected to the roll-up version of the standard, IEEE802.15.4-2011 [7], which therefore replaces the original standards and its amendments [3–9]. This roll-up standard version, and especially its UWB part, is the base for the discussion of this book.

The standardization process did not end to version [7]. As a novelty, the following amendment IEEE802.15.4e-2012 [10] is enhancing and adding new functionalities to the original IEEE 802.15.4 MAC to better support the industrial markets and to permit compatibility with the modifications proposed within the Chinese WPAN standard [8]. For special cases of short-range communications links, the amendment IEEE802.15.4f-2011 [11] provides two new PHY layer approaches, minimum shift keying (MSK) and low rate pulse repetition UWB, that can be used in applications having requirements such as heterogeneous combinations of low-cost, low-energy consumption devices, and thus, a need for very long battery life, reliable communications, precision location, and also reader options. These new PHYs are targeted to applications which need flexibility in the future mass deployments of very dense autonomous active radio frequency identification (RFID) systems globally. Dedicated standard for smart metering applications is presented in the IEEE802.15.4g-2012 [12].

The IEEE802.15.4k, Low Energy Critical Infrastructure Monitoring (LECIM) Interest and Study Group, focused on widespread wireless critical infrastructure monitoring devices and applications, which are operating in a larger framework of IEEE802.15 family. This work ended as a new amendment to the IEEE802.15.4-2011 standard in 2013 [13].

As can be seen, the original IEEE802.15.4 and its successors include several amendments, but this book is focusing on its amendment which is best suitable for exploitation in WBANs. Therefore, the rest of the

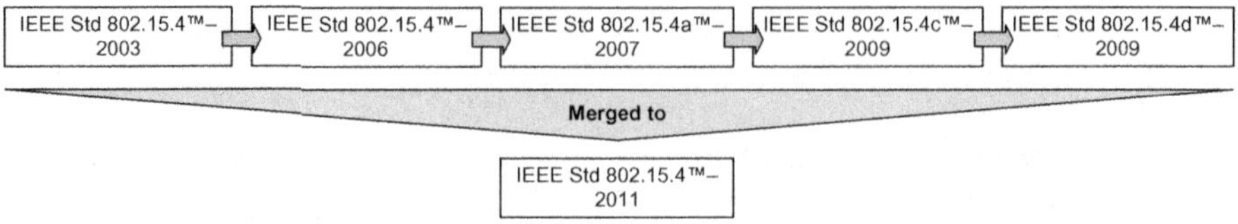


*Figure 2 Evolvement of the IEEE802.15.4 standard before the year 2011.*

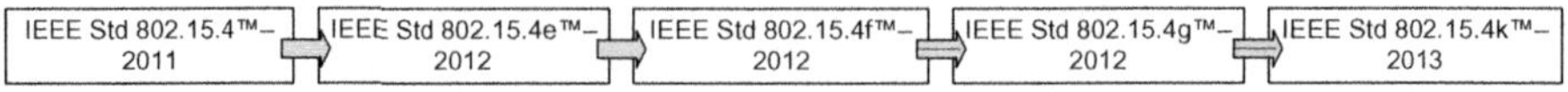


*Figure 3 Evolvement of the IEEE802.15.4-2011 standard after the year 2011.*

book is utilizing the original UWB standard IEEE802.15.4a [5], which is now involved in IEEE802.15.4-2011 [7], as a technology studied in more details. More information on IEEE802.15.4 standard can be found, also as a handbook style, e.g., from Refs. [14,15].

The evolvement of the IEEE802.15.4 standard family is presented in Figure 2 from the beginning to the first roll-up version IEEE802.15.4-2011 and Figure 3 afterward. Currently, there are new amendments for the existing standard in ongoing state.

## 2.1 APPLICATIONS

Numerous applications can rely on the IEEE802.15.4-family based devices. The basic idea behind IEEE802.5.4 was to replace cables using standardized radio technology instead. The main interest in this book is relating to WBANs even though there is a new, dedicated standard IEEE802.15.6-2012 [6] adopted especially for that usage. WBAN is a modern way to convey medical, health, or well-being related data from different parts of a human body through a body access point to end user, services, or remote electronic databases. The potential healthcare related scenario is utilizing small sensors, which are attached to the human body to measure noninvasively different psycho-physiological signs, such as body posture, temperature, respiration rate, and electrocardiograph. Utilizing WBAN solution in human monitoring makes it possible to measure signs simultaneously in different places of a body, and is also allowing mobility for that person. Local access point, i.e., hub, which can be based on several radio standards, such as WiFi base station, cellular base station, Ethernet router, or smartphone, is then transferring the measured information further to the end user, service provider, or medical staff (Figure 1). In fully automated remote monitoring systems, information can be stored to electrical health records from where it is accessed by the authorized persons.

The healthcare related WBAN applications are discussed, e.g., in Refs. [16–19]. These papers cover both the application aspects but also implementation of human body networks, security, energy efficiency and such, making the papers a good starting point for the experimental research viewpoint on WBAN implementation.

Three level hierarchical approach for hospital level networking is proposed in Ref. [20] and wireless hospital architecture, which both are partly utilizing the IEEE802.15.4-2011 standard, are presented, e.g., in [21,22]. The use of WBANs in rehabilitation, occupational healthcare, etc. shortens the time the person is an inpatient at hospital. Other benefits are decreased overall nursing costs and work load of nursing staff which have significant impact on economics of healthcare service providers, hospital districts, and so on.

Based on a star network topology, the network structure itself is very straightforward. In principle, there is no need for, for example, relaying in close body communication links. However, star topology without relay node possibility also limits the quality of the received signal when the communication is based on the sensor nodes in different sides of a human body or during movement, when different parts of the body can block the line-of-sight (LOS) link frequently. Without relaying, the range of the network is limited to one hop, but typically it is enough for most common WBAN applications. The standard supports also peer-to-peer topology, which allows multihops.

Indeed, the final application itself defines the requirement set for communication (e.g., data rate, delay), and especially, to its quality. Different sensors, measurement devices, or other instruments linked to the network have their own communication profiles. This means that each application will occupy the channel differently, their payload varies, and so on. For medical or health related WBAN applications and appliances, there are a varying set of requirements to be met such as security and secrecy issues, low transmission power limit and low specific absorption rate (SAR), small gadget size, and weight. When dealing with carry on devices, energy efficiency is also an important merit directing to low power consumption devices having a long battery life time. For example, in medical applications, fast channel access can also be required for emergency traffic.

Both the technology utilization and application ranges are very wide. Not only health and well-being related applications, the IEEE802.15.4-family can be utilized in industrial, M2M, smart home, smart grid, etc. areas. The possibilities to transfer data from (human) body sensor node toward backbone network using different, heterogeneous solutions, IEEE802.15.4-2011 being just one technology, will be the way how future networks will jointly work.

## 3.1 THE IEEE802.15.4-2011 BASED UWB WBAN

### 3.1.1 The IEEE802.15.4-2011 IR-UWB Properties

In this chapter, the key properties of the IEEE802.15.4-2011 standard for IR-UWB are discussed. The standard documents will introduce the PHY, MAC layer, and security aspects in more details. Shorter overview of the standard can be found, e.g., from Ref. [15]. The following summary of the properties of the IEEE802.15.4-2011 UWB standard is based on Refs. [5,7] if not otherwise referred.

The operational frequencies for IR-UWB are divided into three sub-bands: sub-GHz band at 250–750 MHz, low band at 3.1–5 GHz, and high band at 6–10.6 GHz. For the IR-UWB signal, the standard defines totally 16 channels, all having their own center frequencies ($f_c$). Channel 3 in the low band at $f_c = 4492.8$ MHz and Channel 9 in the high band at $f_c = 7987.2$ MHz are mandatory channels, which an interoperable device needs to support. If the device is operating at sub-GHz band, there is also mandatory band at Channel 0 at $f_c = 499.2$ MHz. The other 13 channels are optional and their utilization is depending on the device vendor. The transmitted signal will occupy 499.2 MHz of the frequency spectrum with few exceptions, namely Channels 4 and 11 have 1331.2 MHz bandwidth, Channel 7 1081.6 MHz, and Channel 15 1354.97 MHz bandwidth [5].

The standard defines star topology for network architecture. Different subnetworks in the neighborhood will be separated using different preambles. The preamble code lengths are either 31 or 127, the former being mandatory for compliant gadgets. The receiver is scanning all the channels to find a beacon and preamble code for channel selection. The channel without traffic is then allocated for communication. Due to this network separation, it is possible to operate several subnetworks in the same environment without any coexistence problem.

Next, the symbol structure defined by the standard is shortly introduced. In the IEEE802.15.4-2011 based UWB systems, the symbol frame is divided into four quarters. The first and the third quarters of the symbol frame are used for transmitting information and the second and the fourth quarters are reserved to act as guarding intervals to decrease intersymbol interference, which is caused by multipath propagation. The severity of the multipath interference depends on the environment. The symbol structure is opened in Figure 4 (redrawn from Ref. [5]). In the figure, $T_{burst}$ and $T_c$ stand for burst and chip lengths, respectively.

The information is modulated using burst position modulation in which the first and the third quarters are used to transmit bit "0" and "1," respectively. Due to the low UWB pulse energy, one burst is formed by $N_{cpb}$ pulses to increase the energy level at reception. The achieved processing gain logarithmically increases with the increasing $N_{cpb}$ but as a drawback, the pulse repetition reduces the achievable data rate.

The standard defines how many slots, length of $T_c$, there can be in one-quarter of a symbol. This number $N$ also limits the maximum number of simultaneous users in a subnetwork who can operate without overlapping during a transmission instant. One-quarter can consist of 2, 8, 16, or 32 time slots, so the system is flexible for different configurations.

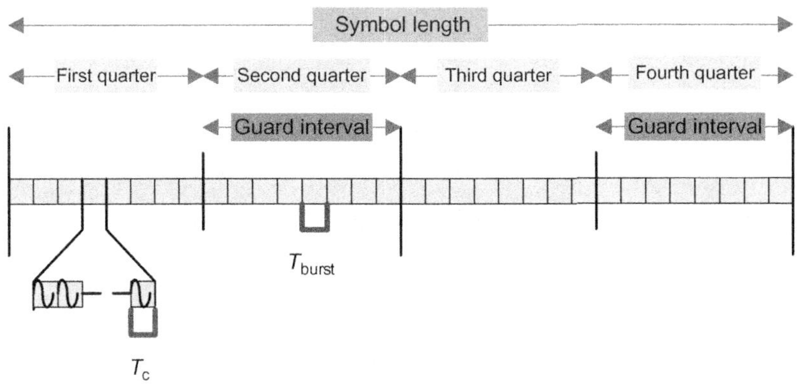


*Figure 4 IEEE802.15.4-2011 UWB symbol structure.*

However, all the nodes involved in the subnetwork are not able to connect with each other. The standard defines full and reduced function devices, FFD and RFD, respectively. The subnetwork needs to have at least one FFD which can communicate also outside of the subnetwork. The simpler nodes, RFD devices, can connect only to FFD node. The main difference in these two category devices is in MAC level: the FFD node is implemented with full MAC services, and the RFD node can support only subset of it.

UWB PHY is also scalable in the means of data rate $R_d$. The number of transmitted pulses, each having length $T_c$, within a burst has direct contribution to the achieved data rate. The standard defines different categories for available data rates from 120 kbps up to 31.2 Mbps. The selected encoded data rate has an impact on the number of pulses used to send one symbol, and thus also to the processing gain available at the reception, finally reflecting to the achievable communications range and system performance. According to the standard, the pulse length is fixed to $T_c = 2$ ns.

Table 1 introduces the scale of symbol rates in [MHz] and the corresponding number of transmitted pulses per burst. The numbers in the table are shown for different numbers of simultaneously nonoverlapping users [7].

**Table 1 Examples of the Symbol Rates and Number of Burst According to the Standard**

| Number of Users | Symbol Rate (MHz) | Number of 2 ns Pulses per Burst |
|---|---|---|
| 2 | 0.12 | 512 |
| | 0.98 | 64 |
| | 7.80 | 8 |
| | 31.20 | 2 |
| 8 | 0.12 | 128 |
| | 0.98 | 16 |
| | 7.80 | 2 |
| | 15.60 | 1 |
| 32 | 0.12 | 32 |
| | 0.98 | 4 |
| | 1.95 | 2 |
| | 3.90 | 1 |

In the IEEE802.15.4-2011, the transmitted UWB signal waveform can be modeled as [5,7]

$$x^{(k)}(t) = \left[1 - 2g_1^{(k)}\right] \sum_{n=1}^{N_{\text{cpb}}} \left[1 - 2s_{n+kN_{\text{cpb}}}\right] \times p\left(t - g_0^{(k)} T_{\text{BPM}} - l^{(k)} T_{\text{burst}} - nT_c\right) \tag{1}$$

where ${g_0}^{(k)}$ and ${g_1}^{(k)}$ are position and phase modulated bits, respectively. $N_{\text{cpb}}$ denotes the number of pulses used to transmit one burst. The scrambling code defining the transmission slot (burst hopping position $l^{(k)}$) within a frame at $k$th transmission interval is presented as $s_{n+kN_{\text{cpb}}} \in \{0, 1\}$, $n = 0, 1, \ldots, N_{\text{cpb}} - 1$. $T_{\text{BPM}}$ is half of the symbol length $T_s$, and it defines the location of the transmitted burst within a frame and thus the transmitted bit. $p(t)$ is the transmitted pulse waveform.

### 3.1.2 Different UWB Receiver Structures for the IEEE802.15.4-2011

The IEEE802.15.4-2011 compatible UWB receivers are supporting both coherent and noncoherent reception approaches. This makes it possible to select either very simple or, alternatively, more complex but reliable implementation for different applications, still remaining the standard compatibility.

The IEEE802.15.4-2011 based transmitted UWB signal is firstly binary burst position modulated (BPM) but then also BPSK modulated. Simply, in the case of bit "0," the burst is transmitted in the first quarter of a symbol frame, as shown in Figure 4. Exact time slot within a quarter is defined by the user dependent random allocation by individual pseudorandom code. Correspondingly in the case of bit "1," the burst is transmitted within a (pseudo) random slot in the third quarter of the frame. Even quarters are used as guard intervals to mitigate the impact of multipath propagation on reception, as illustrated in Figure 4. The use of BPM allows simple noncoherent signal detection for noncritical communication. Independently on the receiver architecture and modulation, the burst position within a frame needs to be detected first.

In the next step, BPSK modulation is used for convolutional channel coded (CC) bits, which can then be utilized in error correction process to improve a packet error rate (PER) at detection. The other channel coding scheme used is Reed–Solomon (RS) coding, and it is utilized by adding redundant bits at the end of the information bit stream. RS coded bits are always position modulated instead of BPSK, which is used with CC. In a data flow, the coding is done as a concatenated code basis, where CC is used as an inner code and RS is forming an outer code. Both RC and CC being systematic codes, the redundant bits are added to the end of the information bit stream. At the end of the channel coding process, the transmitter is producing position modulated information bits and phase modulated parity bits. Depending on the applications' quality requirements, the signal detection can be done either coherently or noncoherently, and thus, this selection is affecting the complexity of the receiver implementation. Noncoherent receiver is not able to utilize CC which is always phase modulated, and thus it is used only by the coherent receiver. To improve system performance, noncoherent receiver can use RS, which is also available for coherent detection. Generally, coherent detection outperforms the noncoherent one in terms of achieved bit error rate (BER) or PER. Naturally, increased complexity is the drawback of more reliable receiver implementation [5,7].

The received signal, which is affected by the propagation channel, can be presented using a generic formulation as

$$r^{(k)}(t) = x^{(k)}(t) * h(t) + n(t) \tag{2}$$

where $r(t)$ and $x(t)$ are received and transmitted burst, respectively, $h(t)$ is channel impulse response, $n(t)$ is white Gaussian noise, and "$*$" denotes convolution.

According to the algorithms presented in the standard [5,7], different receiver types can be identified and classified in several categories by combining coherent and noncoherent detection techniques. The simplest receiver structure is pure energy detector (ED), where detection is based only on the energy comparison between the first and the third quarters of the symbol structure as shown in Figure 4. The decision variable $w_{\mathrm{i}}$ for the ED can be defined as [23]

$$w_{\mathrm{i}}^{(k)} = \int_{q}^{q+T_{\mathrm{burst}}+T_{\mathrm{opt}}} r(t)^2 \,\mathrm{d}t \tag{3}$$

where $q = 2kT_{\mathrm{BPM}} + iT_{\mathrm{BPM}} + h(k)T_{\mathrm{burst}}$ and $T_{\mathrm{opt}}$ is an optimal additional integration time used to benefit the impact of the energy available at the channel due to multipath propagation. Thus, the integration time is the length of a burst and an excess time defined by the length of the channel impulse response containing a significant amount of the desired signal's energy. The performance of an ED can be increased by increasing $T_{\mathrm{opt}}$ if the channel is multipath rich. On the other hand, any extension of $T_{\mathrm{opt}}$ increases also the amount of collected noise at the receiver which means that the selection of $T_{\mathrm{opt}}$ is always a trade-off. By comparing the energies of the symbol quarters 1 and 3, shown in Figure 4, the burst position can be detected by simply comparing the energy levels as

$$w_0^{(k)} \underset{"1"}{\overset{"0"}{\gtrless}} w_1^{(k)} \tag{4}$$

where the energy level of $w_\mathrm{i}$ defines which quarter has more energy, and thus, which bit has been detected. Low signal-to-noise ratio (SNR) decreases the reliability of detection due to the increased proportional amount of noise power being involved in a decision making. The reliability of an ED receiver is significantly better with higher SNR values than in lower SNR cases. Therefore, the use of an ED receiver is recommended for the cases where channel condition is known to be good.

If a coherent detection for BPSK part is obtained after the position detection is carried out, there is a possibility to increase the number of transmitted bits per symbol from one to two. In any case, the burst position needs to be found out before the phase component can be detected. In general, the coherent detection is formulated as

$$v_\mathrm{i}^{(k)} = \int_s^{s+nT_\mathrm{c}} r(t-\tau)b(t)d\tau \tag{5}$$

where locally generated reference burst $b(t)$ is a replica of the transmitted burst which has been affected by the channel impulse response $h(t)$ and can be presented with a priori channel information as

$$b(t) = \left( \sum_{n=1}^{N_{\mathrm{cpb}}} \left[1 - 2s_{n+kN_{\mathrm{cpb}}}\right] \times p(t - nT_\mathrm{c}) \right) * h(t) \tag{6}$$

The third way to implement a noncoherent receiver in the IEEE802.15.4-2011 systems is, so called, bi-orthogonal detection.

In this case, there is a phase modulation included in the transmitted signal but the detection procedure follows the ED. The difference to the simplest ED approach is that now the decision variable is the one coming from the correlation Eq. (5), and the calculation is based on the absolute values. This detection can be formulated as

$$\left|v_0^{(k)}\right| \underset{"1"}{\overset{"0"}{\gtrless}} \left|v_1^{(k)}\right| \tag{7}$$

In addition, it is also possible to detect the phase modulated part of the signal. However, this is done after the position demodulation, according to the decision variable based on Eq. (7), which defines which quarter includes more energy. Finally, the sign of the correlation from Eq. (5) defines the bit value as

$$v_0^{(k)}, v_1^{(k)} \underset{"1"}{\overset{"0"}{\gtrless}} 0 \tag{8}$$

The latter detector can also utilize convolutional decoding, which is embedded via a phase modulation. In this case, the coherent detection is carried out after the noncoherent detection. By increasing the complexity of a receiver, its performance can be improved in terms of BER.

In Refs. [24,25], it is introduced how to increase data rate or number of users in the standard compatible system by utilizing the odd quarters of the symbol frame presented in Figure 4. The adaptation of this new feature requires slight modification to be done in the MAC layer, and this is explained in Ref. [26].

The key idea behind this new approach is to bypass the position based modulation which is utilized in the original standard. By ignoring the requirement of having modulated information either in Q1 or Q3 only, the other, unused quarter can be allocated for another user, or the data rate of the original user can be doubled. As the symbol and frame structure still remains the same as described in the standard, this approach is backward compatible with the devices following the original standard. In addition to increasing either data rate or number of users, this approach allows improvement in system performance by

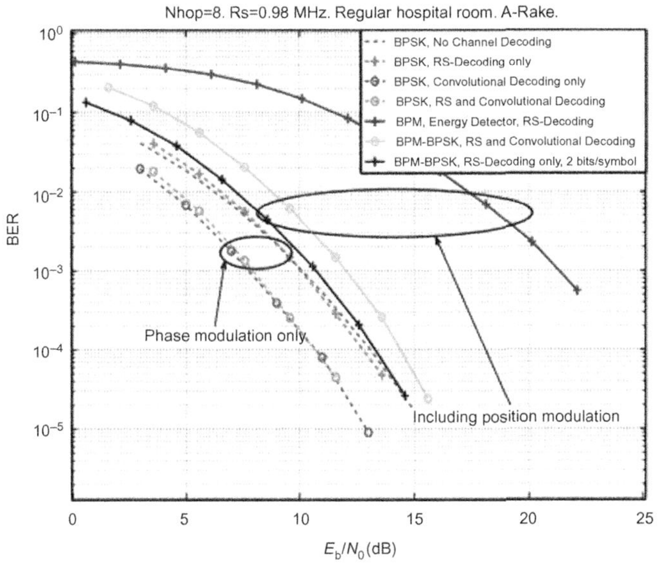


*Figure 5 Performance comparison between the novel and standard based receiver structures.*

introducing more coding gain by jointly transmitting CC bits and bits conveying information. Changes to involve the other quarter need to be done also in the MAC layer, as explained in Ref. [26]. The performance comparison between the system utilizing the new frame structure and the standard based on is shown in Figure 5 [24]. If on–off keying (OOK) is used instead of phase modulation, the modified symbol structure allows also the use of noncoherent receivers.

### 3.1.3 Energy Efficiency

The WBAN standards, such as IEEE802.15.4-2011 or IEEE802.15.6-2012, are designed to consume less energy than, for example, WLAN defined by the different IEEE802.11-family standards. As being a low-energy device, the required energy can be stored for batteries, or it can even be harvested from the environment. Although IEEE802.15.4-2011 is not the most efficient technology in the energy consumption point of view, it is still possible to install devices which are operating for years using the same battery. Nevertheless, any improvements in the energy efficiency are helping to build more efficient networks in the future. If those improvements are not based on PHY modifications, they can be done at the MAC layer. The standard is still tighter for PHY methods, such as the used modulation technique, power control or source and error control coding (ECC), which are strictly described in the standard document. In the MAC layer, it is possible to achieve power efficiency, for example, by using channel listening methods which consumes less energy or by decreasing the retransmissions.

Typically it is a radio in a sensor node which consumes most of the energy; sensors themselves are mainly low power units. One way to improve the energy efficiency is to utilize a cross layer optimization approach in the communication protocol design. For example, via a reliable transmission which utilizes ECC, it is possible to reduce the number of retransmissions, which has a straight impact on the communicating devices' energy consumption. Channel coding can be used to improve the reliability as earlier mentioned. The standard [7] defines channel codes and code rates for the compliant systems. In the IEEE802.15.4-2011 UWB signal, the code rate for RS code with parameters (63,55) is approximately 0.87 and for convolutional code it is 0.5, if channel coding is applied. Thus, the forward error control (FEC) rate is either 0.87 or 0.44, respectively. According to the standard, the use of channel coding is not mandatory.

The ECC is a way to improve the energy efficiency of the communicating system due to the inherent ability to correct certain amounts of erroneous bits. If ECC is activated to meet the performance requirements instead of using retransmissions when errors have detected, the network's energy consumption can be significantly decreased. On the other hand, using different code rates, different system performances are obtained. Another way to improve energy efficiency is to keep overheads at the MAC layer as small as possible, which improves energy efficiency due to the shorter transmitted packet lengths. However, too big changes in the protocol stack will ruin the compatibility and is therefore driven toward proprietary solutions.

Also the channel condition can have impact on the efficiency of error resistance. For example, in Ref. [27], it was shown that in good channel conditions, automatic repeat request (ARQ) is better in an energy-efficiency point of view, while ECC is better when the channel state is bad. Good channel is not causing many bit errors for the transmitted data flow, and therefore ECC is not needed. Again, due to the fact that the receiver can detect transmitted bits mostly without errors, also ARQs are seldom needed. However, if the channel is very bad, ARQs are needed more often, and the use of ECC is then a more energy efficient way to maintain the required performance level. More references on energy efficiency on wireless sensor network context can be found, for example, from Refs. [28,29].

In Ref. [30], the energy efficiency of IEEE802.15.4-2011 based UWB system is studied as a function of different code rate. It can be seen that the code rate from standard does not necessarily give the best energy efficiency. In Figure 6, the energy efficiency is shown for two MAC protocols, time division multiple access (TDMA) and S-Aloha, which are popular channel access schemes in wireless sensor networks [30]. In the presented example, the probability of each node in a S-Aloha approach to compete for channel access is 0.1. The comparative results in Ref. [30] have probabilities, which were ideally defined for each $E_b/N_0$ value. As expected, the energy efficiency improves when the channel condition improves. After a certain $E_b/N_0$, the uncoded transmission is outperforming the coded ones. This point is different for different MAC protocols. In Refs. [5,7], the defined MAC protocols are Aloha and carrier-sense multiple access-collision avoidance (CSMA-CA). CSMA-CA is used in both beacon enabled and nonbeacon enabled personal area networks (PAN). Beacon enabled PAN is utilizing a slotted version of CSMA-CA, while nonbeacon enabled PAN is based on unslotted channel access. The curves present different code rates, 0.87 and 0.97, and uncoded system. As shown, by changing the code rate, it is possible to

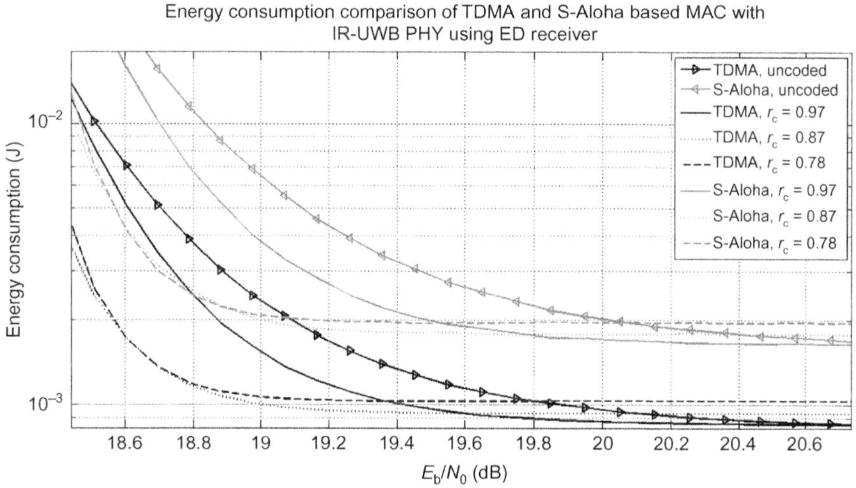


*Figure 6 Energy consumption for TDMA and S-Aloha MAC schemes.*

impact on the system's energy consumption. All the mathematical derivations behind these results can also be found from Refs. [30,31].

## 4.1 WBAN CHANNELS

The vicinity of a human body is a challenging environment for a radio signal propagation point of view. The dielectrical properties of different biological tissues are changing depending on the body structure [32–34] and also on the antenna location. Person's age, gender, size, and so on have an influence on conductivity and permeability of a body, and thus, have an impact on electromagnetic signal propagation around the body. In addition, permittivity and conductivity are frequency dependent, which result in meaningful impact on wideband and UWB communications. Moreover, an LOS link cannot be guaranteed all the time for WBAN links, especially during the instants when humans (or other body structures in general, such as machine and vehicle) are moving. According to, for example, Refs. [35–37], the UWB signal is not propagating through a human body but rather circulates it in a close vicinity of a skin. Therefore, the WBAN links around a body might need a relaying node to pass the signal to the other side, and making the transmission incident more reliable. To allow relaying, the IEEE802.15.4-2011 network needs to be implemented with FFD nodes which can support multihops.

In the following sections, a set of UWB radio channel models which are applicable for the IEEE802.15.4-2011 studies are presented.

### 4.1.1 The IEEE802.15.4a Standard Channel Model

For WBAN use, the IEEE802.15.4a standard has defined radio channel models for system design and testing. These models are described and published in Ref. [38]. Especially, the channel models for frequency range between 2 and 10 GHz are the most interesting in a UWB WBAN viewpoint. The WBAN channel model introduced in Ref. [38] is targeted for the maximum bandwidth of 2 GHz. The model is not valid for wider bandwidths, and smaller bandwidths need to be downfiltered. As pointed out in Ref. [38] the on-body channel model differs from, for example, the indoor channel model due to the different amplitude distributions, path loss characteristics, and envelope of delay profile. In general, it is very important at the transceiver development phase to have as realistic view on the propagation channel characteristics as possible to be able to optimize signal waveform and receiver implementation, and thus, the system performance.

The standard WBAN channel model is based on lognormal amplitude distribution, or in some specific cases, on Nakagami-m distribution. The delay profile, or impulse response, is affected by the profile and structure of the body, as well as the antenna locations on the body. The general feature of the channel impulse response in WBAN environment is heavy clustering around the dominant propagation paths. Longer delays do not contribute significantly in an additional signal energy point of view due to their higher attenuation caused by higher path loss.

The channel models defined in the final report of the IEEE802.15.4a channel model subgroup [39] are based on the modified Saleh–Valenzuela models [40], which were also adopted for high data rate UWB use by the terminated IEEE802.15.3a study group proposal [41]. Other radio channel models applicable for WBAN use are also reviewed in Ref. [42]. Later on in this chapter, some measurement based models are shortly introduced.

In addition to the mathematical formulation of the channel characteristics, the standard channel model document [39] introduces the Matlab™ codes to generate standard compliant radio channel realizations for IEEE802.15.4-2011 studies for different use cases and environments, such as WBAN, office, residential, outdoors, and open areas. Without going into any deeper details on the IEEE802.15.4a channel models, a few examples on the channel characterizations are shown in the following figures: Figure 7 presents the average WBAN power delay profile calculated over 100 channel realizations. Figure 8 shows the number of paths capturing more than 85% of the energy and being attenuated less than 10 dB from the maximum path. Finally, Figure 9 presents the excess and root mean square (RMS) delays. All the parameters are calculated from the same data set, thus 100 generated channel realizations. The Matlab™ code used to generate the examples of the channel realizations can be found from Ref. [39].

As shown in Figure 7, the envelopes of the averaged WBAN channel profiles are monotonically decreasing and there are no strong reflections coming from the longer delays. Due to the very fine resolution of the UWB signal, there are huge amounts of distinguishable paths in the WBAN environment. In the graph, there are presented three

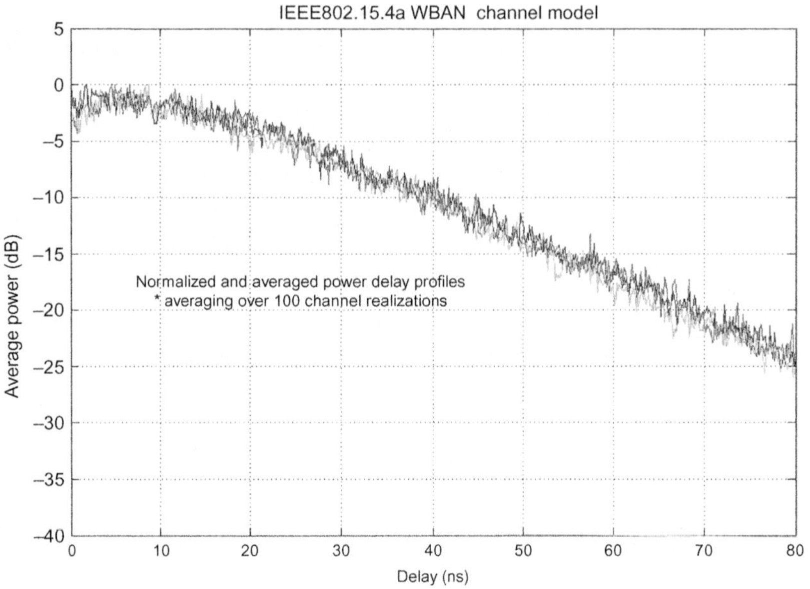


*Figure 7 Average IEEE802.15.4a WBAN power delay profile.*

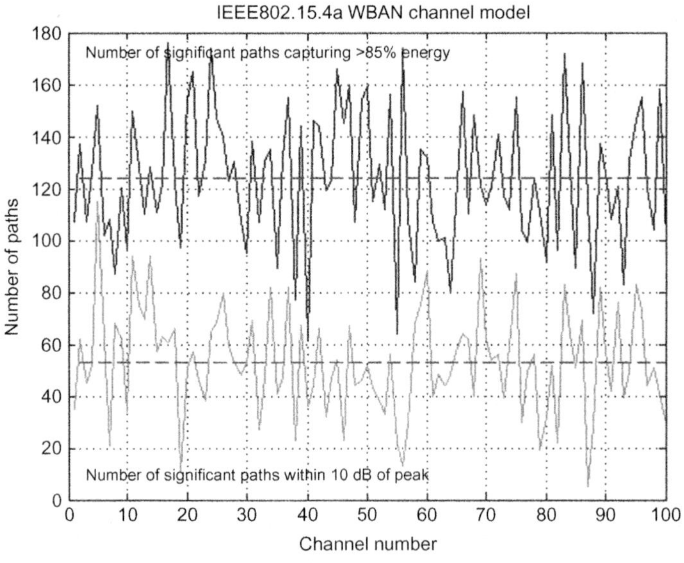


*Figure 8 Number of paths capturing >85% of energy and attenuated less than 10 dB from the peak.*

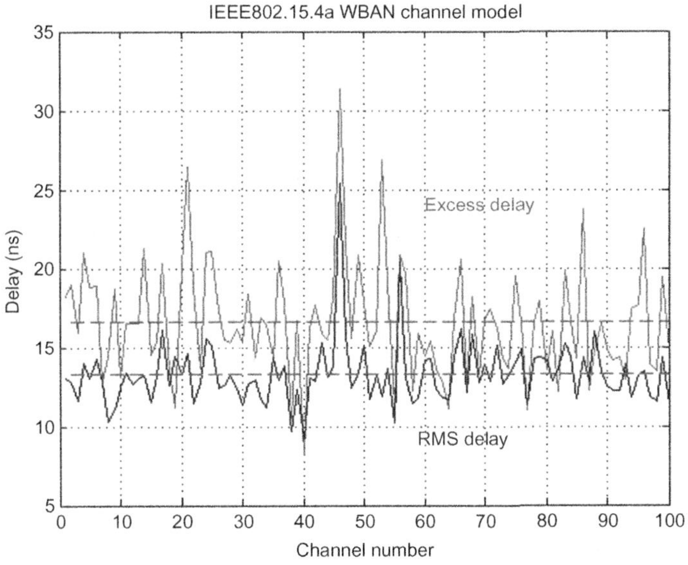


*Figure 9 Excess and RMS delays.*

independent simulations, having 100 independent channel realizations each. The Matlab™ model produces reasonable stable delay profiles, having still some statistical variation between the different realizations.

To tie WBAN applications to a hospital environment, the patient room is the most appropriate place to utilize wireless personal solutions. For example, in Ref. [43], the UWB signal propagation characterization in various distances has been studied in the bedside environment, covering both LOS and non-LOS links originated at a human body and ending to an access point at the pole. Moreover, the WBAN channel characteristics suitable for the IEEE802.15.4-2011 studies have been reported, for example, in Refs. [44,45], which introduce path losses for different WBAN links. In addition, the antenna location and type have an impact on received signal level as pointed out for hip-to-chest and hip-to-wrist links using nine different antennas [46]. Even though all the models are not based on the measurements carried out in a real hospital, they are targeted to model signal propagation in hospital use.

### 4.1.2 Hospital WBAN Channel Model

Other WBAN channel models, which are applicable for the hospital environment, are presented in Refs. [47–54]. These models are based on the WBAN channel measurements carried out at the Oulu

University Hospital, Oulu, Finland. The models describe the WBAN signal propagation characteristics in a real environment of use. The measurements covered several spaces in a hospital, namely regular ward, operation theatre, and hospital corridor. Those environments reflect well the situation where a hospitalized inpatient is most typically using WBAN as a real-time or continuous physiological sign monitoring system. All vital sign measurements do not need to be real time.

The radio channel behind this model was measured within the frequency range of 3.1–10 GHz, thus being an UWB channel and occupying completely the FCC defined UWB band. Due to the frequency domain measurement approach, the environment was kept static during the recordings. Time domain presentations of the channel characteristics have been obtained through inverse fast Fourier transform (IFFT). The corresponding channel model by the IEEE 802.15.6 standard is presented in Ref. [55] and is compared to the model based on the measurements at the Oulu University Hospital in Ref. [56]. Due to the almost 7 GHz frequency band, the empirical model from Ref. [57] has very fine delay resolution, which is much more accurate than the model from Ref. [55].

Figure 10 shows examples of impulse responses representing a typical hospital room at Oulu University Hospital. The antennas are either attached both to the human body, or body and a near pole, which are corresponding on–on body link and on–off body link, respectively. During the measurements, the test person was either standing or lying in a bed. In the on–on body case, the studied link was between sensor nodes located at the middle of the torso and at the left wrist. In the on–off case, the link was measured between a sensor node at a left side of the waist and on a 2 m high pole located 2 m away from the subject. References [47–54] present deeper analysis of the model and measurements.

It should also be noted that the human posture has an impact on the signal propagation. If the environment remains the same, but the human is changing his or her standing position to laying, the channel characteristics are changing, correspondingly. For example, in Ref. [49], the link between a human and a pole is studied with standing and laying positions in a ward, operation theatre, and corridor. The results indicate that even in WBAN links, the measured channel frequency response, and thus also impulse response, is different. In the standing posture, the dynamics of the impulse response is several tens of decibels

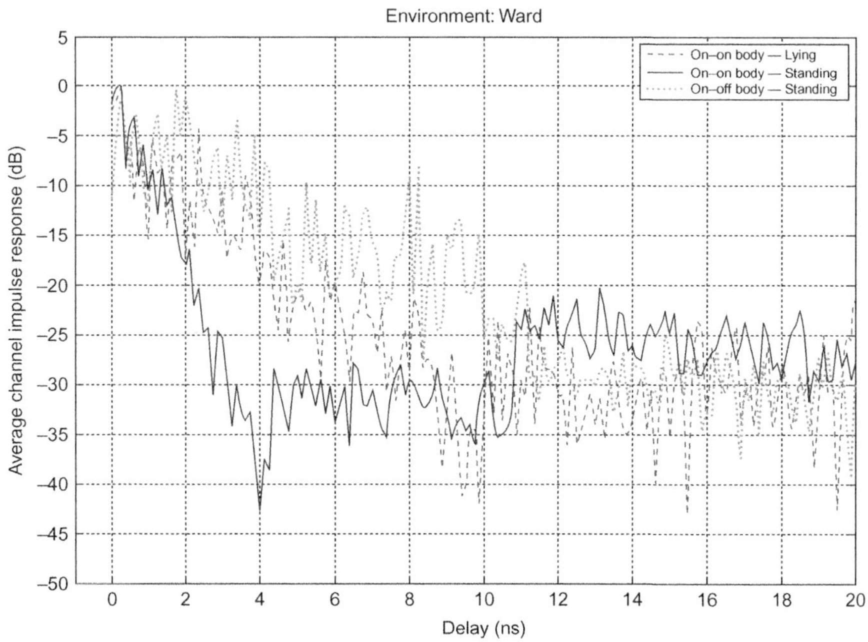


*Figure 10 Impulse responses measured at the hospital ward; on–on and on–off body links.*

higher than in the corresponding laying scenario. In addition, the first detected signal component carries more energy. This difference is basically due to the more obstructed link when the person is in a lying position if compared to the standing position.

### 4.1.3 Other Channel Model Impacts

In Refs. [48,57,58], it has also been reported that implants might have an impact on signal propagation in on-body communication links which are crossing the implant in short distance. Based on the measurements, which also cover the IEEE802.15.4-2011 frequency band, carried out with several test persons, a radio link which crosses an aortic valve implant has significant difference in the maximum power level of the measured channel impulse response if compared to the cases with persons who do not have an implant.

As an example, Figure 11 [58] presents three measurement based channel impulse responses of on-body link for three different persons. One of the test persons had a titanium alloy aortic valve implant and the presented impulse responses indicated that the implant really effects on the channel behavior. This phenomenon has also been

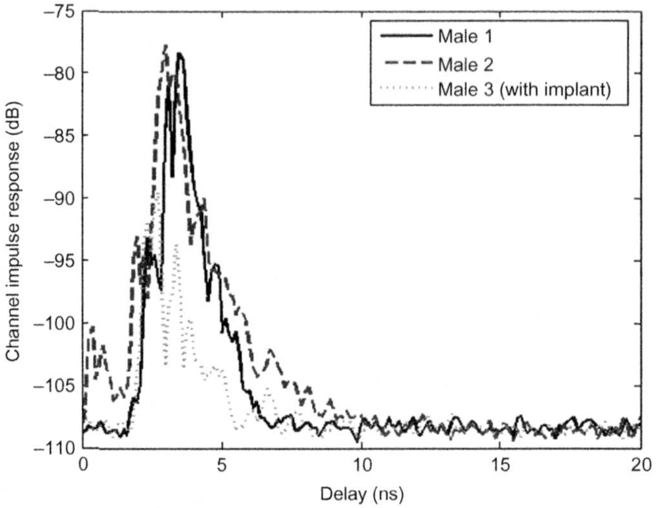


*Figure 11 Measured impulse responses for on-body links. One of the three test persons had an aortic valve implant.*

demonstrated by computer simulations using the 3D human model in an immersive visualization environment [59]. The modeling in the simulation environment was carried out by the National Institute of Standards and Technology (NIST) in Gaithersburg, MD, USA. The simulated links are modeled using partially the same parameters and antenna positions than the experiment in Refs. [48,57,58] utilized. Due to the calculation demands, the simulated frequency band in Ref. [59] was narrower than what was used in the measurements. Also the antenna characteristics were different. However, a similar type of behavior in the channel characteristics was found when the radio link cross the implant. These simulation results and their comparison to the measurement based results are reported in Ref. [59] in more details.

## 5.1 EXAMPLES OF PERFORMANCE ANALYSIS FOR DIFFERENT RECEIVER STRUCTURES

The IEEE802.15.4-2011 based systems have been studied for various applications; healthcare and wellness are just part of the whole utilization portfolio. In this chapter, some examples of the comparative analysis of the IEE802.15.4-2011 based receivers' performances applied to WBAN communication are presented. The original results are presented in Refs. [23,60–69], from where the deeper analysis can also be found.

The receiver structures studied here are based on the ones introduced in Chapter 3, and are all following the standard [5,7]. Again, the signal propagation near body is notified by using proper channel models to give a realistic close to human signal model for reception, and therefore the performance comparison is relevant for WBAN applications.

The results presented in this chapter are based on the Matlab™ simulations. The generated simulation model follows the IEEE802.15.4-2011 standard; thus the implemented signal and receiver models are compliant with the standard, and the presented results are comparable with other related results. The channel model typically used is the one generated based on the measurements carried out at the Oulu University Hospital. The receiver performances are presented as a function of SNR ($E_b/N_0$) if not otherwise stated. The energy of one bit, $E_b$, contains the signal energy included in one burst (see Figure 4) and $N_0$ represents zero mean Gaussian noise. To get reliable simulation results, $10^6$ bits per each $E_b/N_0$ value are transmitted through the channel.

As shown in Ref. [64], the energy detection receiver is very sensitive to the symbol rate in additive white Gaussian noise (AWGN) channel. The higher the $R_d$, the better is the performance. In ED type detector, lower $R_d$ increases the amount of noise as compared to signal energy at the detection process. The difference to noncoherent position detection, which utilizes convolutional coding is 3–10 dB depending on the $R_d$, using $N = 2$. When changing the channel model from AWGN to the measurement based WBAN model described in Ref. [49], the detrimental performance degradation can be seen, but the mutual difference between the modulation methods will be smaller, around 5 dB. In addition, convolutional coding will improve the performance of the noncoherent receiver by around 2 dB. The comparative results in terms of BER vs. $E_b/N_0$ are presented in Figure 12 for AWGN and in Figure 13 for the WBAN channel [64].

When using the ED receiver, its performance can be adjusted by proper selection of integration time, e.g., the time a receiver listens to a channel and collects energy to be used at the detection process. In the AWGN channel, the integration time is the one of the burst length due to the fact that the channel does not introduce any replicas of the transmitted signal, thus, no multipath propagation exists. However, in the case of the multipath channel, an ED receiver can collect energy and utilize the energy from the replicas arriving shortly later than the

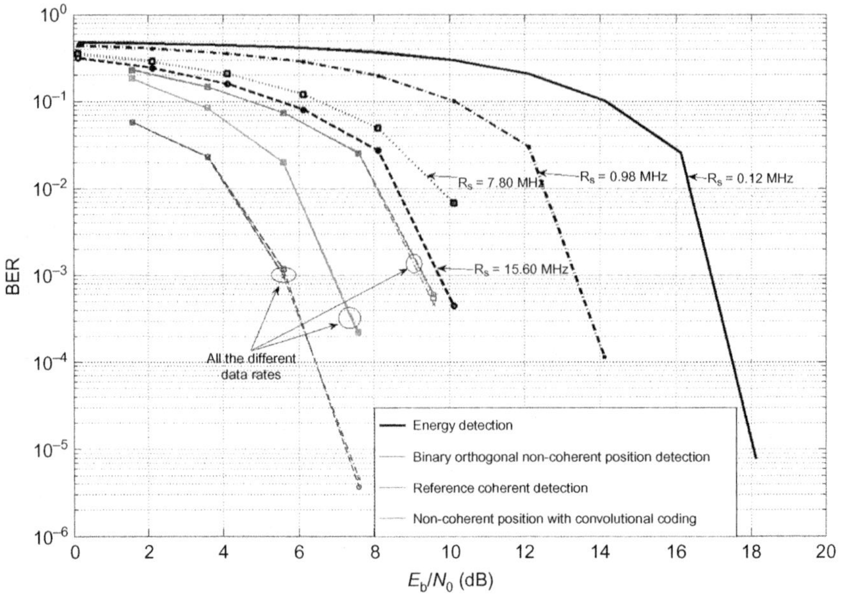


*Figure 12 The impact of data rate on the performance on different receiver implementations in a AWGN channel.*

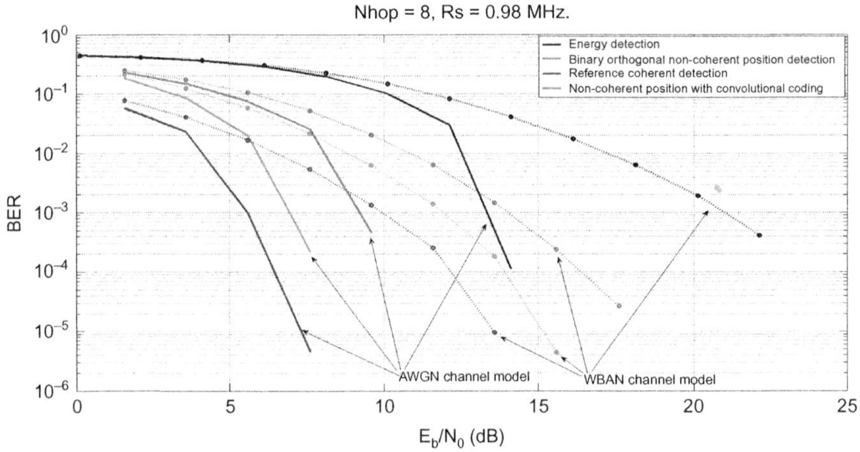


*Figure 13 Comparison of the channel model to the performances of different receiver structures.*

first arriving signal component. The length of this extension depends on the channel characteristics. In an optimal case, the receiver can adjust the time according to the channel condition. This topic has been discussed, for example, in Refs. [23,60,61,68,69].

Using the same simulation model, the impact of utilizing the rake receiver approach has been studied, for example, in Ref. [65]. Again, the channel model used in Ref. [65] is from Ref. [49]. The studied approaches were partial and selective rake receivers, P-rake and S-rake, respectively. Different rake receivers are differently taking advantage of the possible multipath propagated signal energy. P-rake will utilize the $M$ first arrived paths, independently on their strength, whereas S-rake is capturing $M$ the strongest paths [70–72]. In the latter case, delay estimation is required, which means increased complexity in the receiver implementation. Of course, both rake implementations require phase and amplitude estimation in combining process at the detection. A reference model shown in the figure utilizes all the signal energy available in the channel and the receiver is called all rake, A-rake [71,72]. However, from the implementation point of view, A-rake is just a theoretical representation due to the huge amount of fingers it needs to have, and thus, also high computational power is also required. Particularly in UWB solutions, such as the IEEE802.15.4-2011 devices are, the different multipath components can be distinguished with high accuracy, and thus, there are lots of distinguishable signal components available for detection purposes. In the case, where signal bandwidth is much smaller than in UWB, and correspondingly, the delay resolution is coarser, also A-rake receiver can be feasible. In that case, the multipath propagated signal energy is clustered to wider individual path components in the delay domain (called delay bins) and there are not many of them available.

Figure 14 shows BER performances of different rake receivers as a function on $E_b/N_0$ [65]. Symbol rate is fixed to $R_s = 0.98$ MHz. The A-rake receiver is used as a reference to highlight the maximum achievable performance the studied receiver can obtain. Increasing the number of S-rake branches improves the system performance to some point. However, with the increased number of rake branches, the energy contribution of individual paths will decrease, and the performance will saturate due to the characteristics of the channel delay profile. Typically this is the case because the signal energy is attenuating gradually, and in the delays long enough, there are no additional energies, just noise, as illustrated, for example, in Figures 7 and 10. Corresponding performance for the P-rake can be seen in Figure 15, originally also presented in Ref. [65]. The trend of P-rake's performance follows the performance of S-rake's. However, to achieve the same performance level, measured

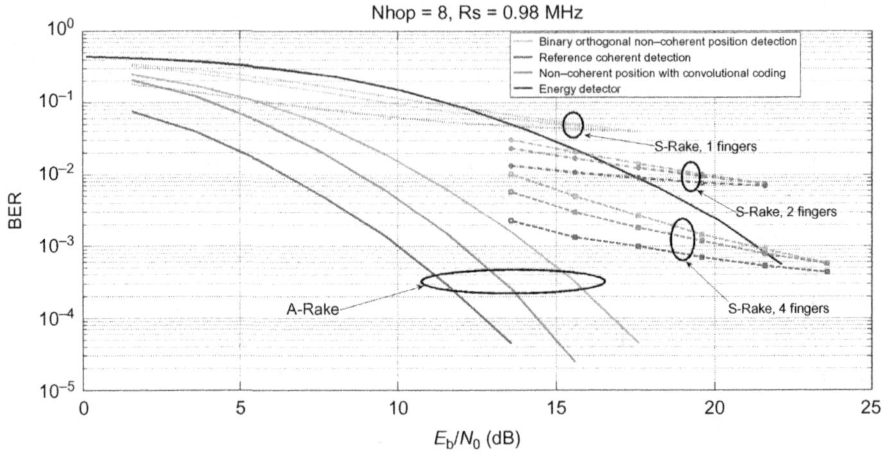


*Figure 14 Performance comparison of IEEE802.15.4-2011 compatible A- and S-rake receivers in WBAN channel.*

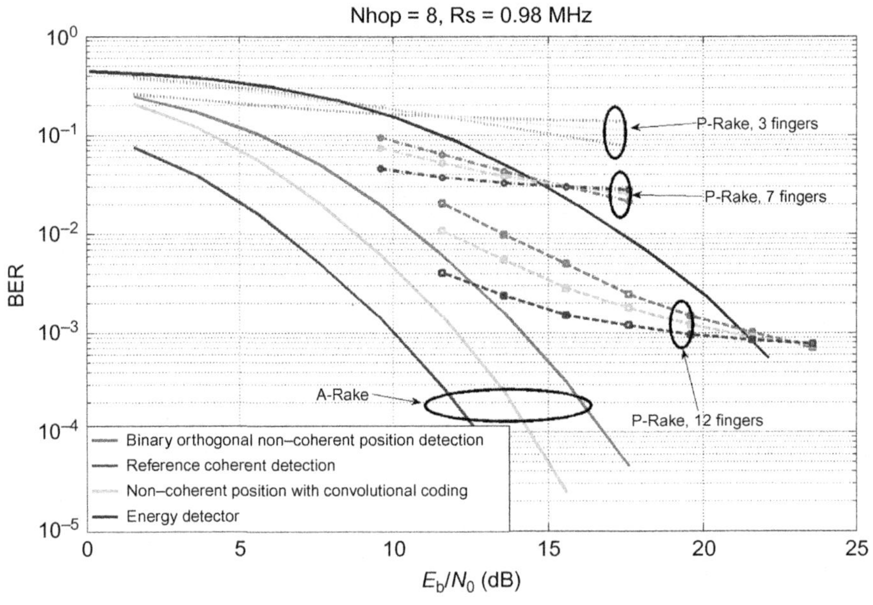


*Figure 15 Performance comparison of IEEE802.15.4-2011 compatible A- and P-rake receivers in WBAN channel.*

by BER, the P-rake receiver requires more fingers thus being a suboptimal solution. If targeting to optimal receiver implementation, unpractical A-rake is required. Then, all the energy the channel can provide through the multipath propagation is utilized at the reception.

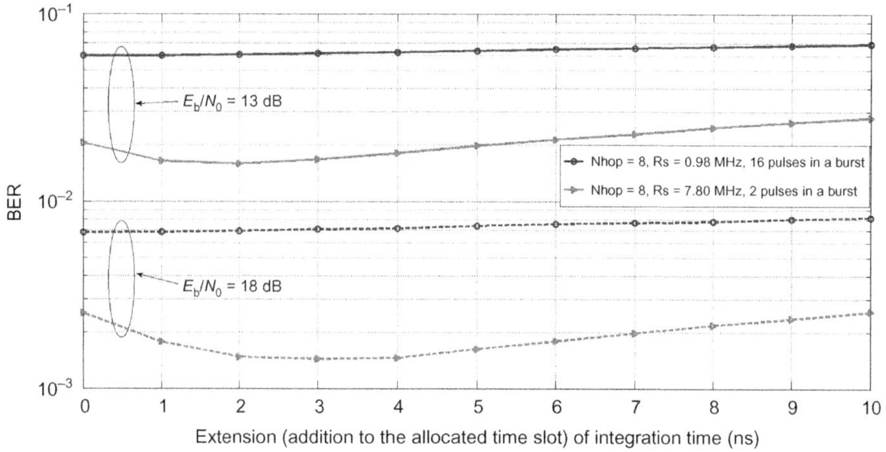


*Figure 16 Impact of integration time extension to an ED receiver's performance.*

In the case of an ED receiver, in the performance shown in Figures 14 and 15, the detection is done based on the energy coming from one finger only because an ED detector is not able to distinguish consecutive multipath components. On the other hand, the energy is integrated longer than the burst interval to collect energy from several multipath propagated signal components, as shown in Eq. (3). The optimal extension for integration time depends on how multipath rich the channel is, and what the SNR is at the reception. The impact of integration time extension to the ED receiver's performance is shown in Figure 16 [65].

The performance of an ED receiver is also sensitive to data rate. In the case of low data rate (when long burst length is used), the ED receiver collects more noise than the coherent receiver approaches in the corresponding situation, and its performance is worse than when high data rates are used. The figure impresses this observation with two data rates and $E_b/N_0$ values. If the $E_b/N_0$ value is high enough, there can be significant energy available at the channel which can be used at the reception and the achieved BER is good. However, with small $E_b/N_0$ values, the received energy is mainly noise, and the receiver performance is therefore bad.

Moreover, the impact of different channel models and data rate (i.e., burst length) for ED and rake receivers' performance is studied in Ref. [69]. The WBAN channel models used are based on the IEEE802.15.6 standard [73] and the one from Ref. [52]. Both channel models are compatible with the IEEE802.15.4-2011 signal model. As a whole, it is evident that the receivers selection problem is multifold due to many different effecting issues. Therefore, careful case study investigations are needed. This always requires understanding of the radio channel, too.

## 6.1 CONCLUSION

The IEEE802.15.4-2011 standard defined a wireless, short-range technology that can be utilized in different kinds of applications. The PHY studied in this book is based on IR-UWB technology. WBANs targeted for healthcare and welfare applications are one interesting implementation area where the standard can successfully be exploited. However, the background of the standard development was in M2M type connections, which makes the standard based solutions widely recoverable. Nowadays, the technology is matured and there is plenty of research behind the current state of the art.

This short book presented briefly the performance comparisons between the different receiver structures, specified by the standard, for human body area communication used in healthcare and welfare applications. In addition, the radio channel models that can be utilized in WBAN with IEEE802.15.4-2011 standard compatible devices were summarized.

According to the standard, it is possible to make a trade-off between a simple, but low transmission quality receiver implementation and more complex receivers, which give better performance, with higher cost. The simplest receiver implementations are based on pure energy detection while the more sophisticated receivers are utilizing coherent detection and rake receivers. Using different standard based receiver implementation, the overall system performance can vary significantly.

The future WBAN solutions will definitely utilize the newer version of the IEEE standard, namely IEEE802.15.6-2012, which is dedicated for body area communication, both for medical and commercial and entertainment applications. Also this new standard is using UWB, IR as a mandatory PHY technology, and the results based on the work presented in this book can partially be utilized. Also the radio channel models introduced are valid for the newer standard.

## REFERENCES

[1] IEEE Standard for Information technology—Telecommunications and information exchange between systems local and metropolitan area networks—Specific requirements Part 11: Wireless LAN Medium Access Control (MAC) and Physical Layer (PHY) Specifications. IEEE Std 802.11™-2012, IEEE Computer Society, 2793 pp.

[2] IEEE Standard for Information technology—Telecommunications and information exchange between systems local and metropolitan area networks— Part 15.1: Wireless Medium Access Control (MAC) and Physical Layer (PHY) Specifications for Wireless Personal Area Networks (WPANs). IEEE Std 802.15.1™, IEEE Computer Society, 600 pp.

[3] IEEE Standard for local and metropolitan area networks—Part 15.4: Wireless Medium Access Control (MAC) and Physical Layer (PHY) Specifications for Low-Rate Wireless Personal Area Networks (LR-WPANs). IEEE Std 802.15.4™-2003, IEEE Computer Society, 679 pp.

[4] IEEE Standard for local and metropolitan area networks—Part 15.4: Wireless Medium Access Control (MAC) and Physical Layer (PHY) Specifications for Low-Rate Wireless Personal Area Networks (WPANs). IEEE Std 802.15.4™-2006, IEEE Computer Society, 323 pp.

[5] IEEE Standard for local and metropolitan area networks—Part 15.4: Wireless Medium Access Control (MAC) and Physical Layer (PHY) Specifications for Low-Rate Wireless Personal Area Networks (WPANs): Amendment 1: Add Alternate PHYs. IEEE Std 802.15.4a™-2007, IEEE Computer Society, 203 pp.

[6] IEEE Standard for local and metropolitan area networks—Part 15.6: Wireless Body Area Networks. IEEE Std 802.15.6™-2012, IEEE Computer Society, 271 pp.

[7] IEEE Standard for local and metropolitan area networks—Part 15.4: Part 15.4: Low-Rate Wireless Personal Area Networks (LR-WPANs). IEEE Std 802.15.4™-2011, IEEE Computer Society, 314 pp.

[8] IEEE Standard for local and metropolitan area networks—Part 15.4: Wireless Medium Access Control (MAC) and Physical Layer (PHY) Specifications for Low-Rate Wireless Personal Area Networks (WPANs): Amendment 2: Alternative Physical Layer Extension to support one or more of the Chinese 314-316 MHz, 430-434 MHz, and 779-787 MHz bands. IEEE 802.15.4c™-2009, IEEE Computer Society, 33 pp.

[9] IEEE Standard for local and metropolitan area networks—Part 15.4: Wireless Medium Access Control (MAC) and Physical Layer (PHY) Specifications for Low-Rate Wireless Personal Area Networks (WPANs): Amendment 3: Alternative Physical Layer Extension to support the Japanese 950 MHz bands. IEEE Std 802.15.4d™-2009, IEEE Computer Society, 39 pp.

[10] IEEE Standard for local and metropolitan area networks—Part 15.4: Low-Rate Wireless Personal Area Networks (WPANs): Amendment 1: Amendment to the MAC sub-layer. IEEE 802.15.4e™-2012, IEEE Computer Society, 225 pp.

[11] IEEE Standard for local and metropolitan area networks—Part 15.4: Low-Rate Wireless Personal Area Networks (WPANs): Amendment 2: Active Radio Frequency Identification (RFID) System Physical Layer (PHY). IEEE P802.15.4f™-2012, IEEE Computer Society, 72 pp.

[12] IEEE Standard for local and metropolitan area networks—Part 15.4: Low-Rate Wireless Personal Area Networks (LR-WPANs) Amendment 3: Physical Layer (PHY) Specifications for Low-Data-Rate, Wireless, Smart Metering Utility Networks. IEEE P802.15.4g™-2012, IEEE Computer Society, 252 pp.

[13] IEEE Standard for Local and metropolitan area networks—Part 15.4: Low-Rate Wireless Personal Area Networks (LR-WPANs)—Amendment 5: Physical Layer Specifications for Low Energy, Critical Infrastructure Monitoring Networks. IEEE P802.15.4k™ – 2013, IEEE Computer Society, 149 pp.

[14] Gutièrrez J, Winkel L, Callaway Jr E, Barrett Jr R. Low-rate wireless personal area networks. Enabling wireless sensors with IEEE 802.15.4™ Standards Information Network. New York, NY: IEEE Press; 2010, 245 pp.

[15] Karapistoli E, Pavlidou F-N, Gragopoulus I, Tsetsinas I. An overview of the IEEE 802.15.4a standard. IEEE Commun Mag 2010;48(1):47–53.

[16] Jovanov E. Wireless technology and system integration in body area networks for m-Health applications. In: The twenty-seventh annual international conference of the IEEE engineering in medicine and biology society, Shanghai, China; 2005.

[17] Otto C, Milenkovic A, Sanders C, Jovanov E. System architecture of a wireless body area sensor network for ubiquitous health monitoring. J Mobile Multimedia 2006;1(4):307–26.

[18] Milenkovic A, Otto C, Jovanov E. Wireless sensor networks for personal health monitoring: issues and an implementation. Comput Commun 2006;29(13–14):2521–33.

[19] Jovanov E, Milenkovic A. Body area networks for ubiquitous healthcare applications: opportunities and challenges. J Med Syst 2011;35(5):1245–54.

[20] Ben Slimane J, Song Y-Q, Koubâa A, Frikha M. A three-tiered architecture for large-scale wireless hospital sensor networks. In: The second international joint conference on biomedical engineering systems and technologies, Porto, Portugal; 2009.

[21] Hämäläinen M, Pirinen P, Shelby Z, Iinatti J. Wireless applications in healthcare and welfare. In: Frigyes I, Bito J, Bakki P, editors. Chapter in advances in mobile and wireless communications: views of the 16th IST mobile and wireless communication summit, Series: Lecture notes in electrical engineering, vol. 16. 2008, XXVII, 407 pp, ISBN: 978-3-540-79040-2.

[22] Ashraf MI, Härkönen M, Hämäläinen M, Riekki J. Health care process management supported by wireless technology. In: Finnish signal processing symposium, Oulu, Finland; 2007.

[23] Niemelä V, Hämäläinen M, Iinatti J. IEEE 802.15.4a UWB receivers in medical applications. Int J Ultra Wideband Commun Syst 2011;2(2):73–82. Available from: http://dx.doi.org/10.1504/IJUWBCS.2011.044599 InderScience Publishers, ISSN: 1758-728X (Print), ISSN: 1758-7298 (Online).

[24] Niemelä V, Hämäläinen M, Iinatti J. Improved usage of time slots of the IEEE 802.15.4a UWB system model. In: The second international workshop on future wellness and medical ICT systems in conjunction with the fourteenth international symposium on wireless personal multimedia communications, Brest, France; 2011.

[25] Niemelä V, Hämäläinen M, Iinatti J. Data transmission in ultra wideband. Patent Pending, Application No. FI20115956; 2011.

[26] Paso T. Informing of modulation method. Patent Pending, Application No. FI20115959; 2011.

[27] Agarwal R, Popovici E, Feo O, O'Flynn B. Energy driven choice of error recovery protocols in embedded sensor network systems. In: The IEEE international conference on sensor technologies and applications, Valencia, Spain; 2007.

[28] Vuran M, Akyildiz I. Error control in wireless sensor networks: a cross layer analysis. IEEE/ACM Trans Network 2009;17(4):1186–99.

[29] Vuran M, Akyildiz I. XLP: a cross-layer protocol for efficient communication in wireless sensor networks. IEEE Trans Mobile Comput 2010;9(11):1578–91.

[30] Karvonen H, Pomalaza-Raez C, Hämäläinen M. A cross-layer optimization approach for lower layers of the protocol stack in sensor networks. ACM Trans Sens Network 2014 [in press].

[31] Karvonen H, Iinatti J, Hämäläinen M. A cross-layer energy-efficiency optimization model for WBAN using IR-UWB transceivers. Springer Telecommunication Systems (TSMJ), special issue on Research advances in energy-efficient MAC protocols for WBANs [accepted for publication].

[32] Gabriel C, Gabriely S, Corthout E. The dielectric properties of biological tissues: I. Literature survey. Phys Med Biol 1996;41:2231–49.

[33] Miklavcic D, Pavselj N, Hart F. Electric properties of tissues. Wiley encyclopedia of biomedical engineering. John Wiley & Sons; Hoboken, NJ, USA: 2006, 12 p.

[34] Wegmueller M, Kuhn A, Froehlich J, Oberle M, Felber N, Kuster N, et al. An attempt to model the human body as a communication channel. IEEE Trans Biomed Eng 2007;54 (10):1851–7.

[35] Fort A, Desset C, Ryckaert J, De Doncker P, Van Biesen L, Donnay S. Ultra wide-band body area channel model. In: IEEE international conference on communications, Soul, Korea; 2005.

[36] Fort A, Desset C, De Doncker P, Wambacq P, Van Biesen L. An ultra-wideband body area propagation channel model—from statistics to implementation. IEEE Trans Microw Theory Tech 2006;54(4):1820–6.

[37] Zasowski T, Meyer G, Althaus F, Wittneben A. Propagation Effects in UWB Body Area Networks. In: The IEEE international conference on ultra-wideband, Zurich, Switzerland; 2005.

[38] Molisch AF, Cassioli D, Chong C-C, Emami S, Fort A, Kannan B, et al. A comprehensive standardized model for ultrawideband propagation channels. IEEE Trans Antenn Propag 2006;54(11).

[39] Molisch A, Balakrishnan K, Cassioli D, Chong C, Emami S, Fort A, et al. IEEE 802.15.4a channel model—final report; 2004.

[40] Saleh A, Valenzuela R. A statistical model for indoor multipath propagation. IEEE J Sel Area Commun 1987;5(2):128–37.

[41] Foerster J. Channel modeling sub-committee, Final report. IEEE P802.15-02/490r1-SG3a, 2003, 52 pp.

[42] Taparugssanagorn A, Rabbachin A, Hämäläinen M, Saloranta J, Iinatti J. A review of channel modelling for wireless body area network in wireless medical communications. In: The eleventh international symposium on wireless personal multimedia communications, Saariselkä, Finland; 2008.

[43] Taguchi Y, Kumazawa M, Takizawa K, Ikegami T. Evaluation of wideband channel for bedside UWB radio link. In: The third international symposium on medical information and communication technology, Montreal, Canada; 2009.

[44] Kumpuniemi T, Tuovinen T, Hämäläinen M, Yekeh Yazdandoost K, Vuohtoniemi R, Iinatti J. Measurement-based on-body path loss modelling for UWB WBAN communications. In: The seventh international symposium on medical information and communication technology, Tokyo, Japan; 2013.

[45] Haneda K, Takada J-I, Takizawa K-I, Vainikainen P. A delay domain pathloss model for ultra wideband ranging. In: IEEE antennas and propagation society international symposium, Charleston, SC; 2009.

[46] Roblin C, Sibille A. Modeling of the influence of body-worn antennas upon the path loss variability in UWB WBAN scenarios. In: 2011 XXXth URSI general assembly and scientific symposium, Istanbul, Turkey; 2011.

[47] Taparugssanagorn A, Pomalaza-Ráez C, Isola A, Tesi R, Rabbachin A, Hämäläinen M, et al. UWB channel modelling for wireless body area networks in medical applications. In:

The third international symposium on medical information and communication technology, Montreal, Canada; 2009.

[48] Taparugssanagorn A, Pomalaza-Raez C, Tesi R, Hämäläinen M, Iinatti J. Effect of body motion and the type of antenna on the measured UWB channel characteristics in medical applications of wireless body area networks. In: IEEE international conference on ultra wideband, Vancouver, Canada; 2009.

[49] Taparugssanagorn A, Pomalaza-Ráez C, Tesi R, Isola A, Hämäläinen M, Iinatti J. UWB channel for wireless body area networks in a hospital environment. In: The twelfth international symposium on wireless personal multimedia communications, Sendai, Japan; 2009.

[50] Taparugssanagorn A, Zhen B, Tesi R, Hämäläinen M, Iinatti J, Kohno R. A dynamic channel model of UWB-WBAN for some medical applications. In: The fourth international symposium on medical information and communications technology, Taipei; Taiwan, 2010.

[51] Taparugssanagorn A, Pomalaza-Raez C, Hämäläinen M, Tesi R, Iinatti J, Kohno R. UWB channel characteristics in the proximity of a dynamic human body for WBAN medical applications. In: The fourth international symposium on medical information and communications technology, Taipei, Taiwan; 2010.

[52] Taparugssanagorn A, Pomalaza-Raez C, Isola A, Tesi R, Hämäläinen M, Iinatti J. UWB channel modelling for wireless body area networks in a hospital. Int J Ultra Wideband Commun Syst 2010;1(4):226–36. Available from: http://dx.doi.org/10.1504/IJUWBCS.2010.034304 InderScience Publishers.

[53] Taparugssanagorn A, Zhen B, Tesi R, Hämäläinen M, Iinatti J, Kohno R. A UWB WBAN channel model based on a pseudo-dynamic measurement. SpringerLink, Ann Telecomm 2010;9. Available from: http://dx.doi.org/10.1007/s12243-010-0228-5.

[54] Taparugssanagorn A, Pomalaza-Ráez C, Isola A, Tesi R, Hämäläinen M, Iinatti J. Preliminary UWB channel study for wireless body area networks in medical applications. Int J Ultra Wideband Commun Syst 2011;2(1):14–22. Available from: http://dx.doi.org/10.1504/IJUWBCS.2011.038632 InderScience Publishers.

[55] Yazdandoost KY, Sayrafian-Pour K. Channel model for body area network (BAN). doc: IEEE P802.15-08-0780-12-0006; 2010

[56] Viittala H, Hämäläinen M, Iinatti J. Impact of difference in WBAN channel models on UWB system performance. In: 2010 international symposium on information theory and its applications and 2010 international symposium on spread spectrum techniques and applications, Taichung, Taiwan; 2010.

[57] Hämäläinen M, Taparugssanagorn A, Tesi R, Iinatti J. Wireless medical communications using UWB. In: 2009 IEEE international conference on ultra wideband, Vancouver, Canada; 2009.

[58] Hämäläinen M, Taparugssanagorn A, Iinatti J. On the WBAN radio channel modelling for medical applications. In: The fifth European conference on antennas and propagation, Roma, Italy; 2011.

[59] Yang W-B, Sayrafian-Pour K, Hagedorn J, Terrill J, Yazdandoost KY, Taparugssanagorn A, et al. Impact of an aortic valve implant on body surface UWB propagation: a preliminary study. In: The fifth international symposium on medical information and communication technology, Montreux, Switzerland; 2011.

[60] Niemelä V, Rabbachin A, Taparugssanagorn A, Hämäläinen M, Iinatti J. A comparison of UWB WBAN receivers in different measured hospital environments. In: The third international symposium on applied sciences in biomedical and communication technologies, Rome, Italy; 2010.

[61] Niemelä V, Haapola J, Hämäläinen M, Iinatti J. Integration interval and threshold evaluation for an energy detector receiver with PPM and OOK modulations. In: Workshop on ultra wide band for body area networking (UWBAN) in conjunction with seventh international conference on body area networks, Oslo, Norway; 2012.

[62] Särestöniemi M, Tuovinen T, Hämäläinen M, Yazdandoost K, Kaivanto E, Iinatti J. Applicability of finite integration technique for the modelling of UWB channel characteristics. In: The sixth international symposium on medical information and communication technology, San Diego, CA, USA; 2012.

[63] Särestöniemi M, Tuovinen T, Hämäläinen M, Yazdandoost K, Iinatti J. Channel modeling for UWB WBAN on-off body communication link with finite integration technique. In: Workshop on ultra wide band for body area networking (UWBAN) in conjunction with seventh international conference on body area networks, Oslo, Norway; 2012.

[64] Niemelä V, Rabbachin A, Taparugssanagorn A, Hämäläinen M, Iinatti J. Performance of IEEE 802.15.4a UWB WBAN receivers in a real hospital environment. In: The fourth international symposium on medical information and communications technology, Taipei, Taiwan; 2010.

[65] Niemelä V, Iinatti J, Hämäläinen M, Taparugssanagorn A. On the energy detector, P- and S-rake receivers in a measured UWB channel inside a hospital. In: The third international symposium on applied sciences in biomedical and communication technologies, Rome, Italy; 2010.

[66] Niemelä V, Hämäläinen M, Iinatti J, Taparugssanagorn A. P-rake receivers in different measured WBAN hospital channels. In: The fifth international symposium on medical information and communication technology, Montreux, Switzerland; 2011.

[67] Hämäläinen M, Niemelä V, Iinatti J, Kohno R. Performance comparison of the different IR-UWB receivers in wireless body area networks. In: 2011 IEEE international conference on ultra-wideband, Bologna, Italy; 2011.

[68] Niemelä V, Hämäläinen M, Iinatti J, Kohno R. IEEE 802.15.4a UWB receivers' performance in different body area network channels. In: The fourteenth international symposium on applied sciences in biomedical and communication technologies, Barcelona, Spain; 2011.

[69] Hämäläinen M, Niemelä V, Iinatti J, Kohno R. Performance of UWB receivers using IEEE802.15.4a's minimum burst lengths. In: The sixth international symposium on medical information and communication technology, San Diego, CA; 2012.

[70] Cassioli D, Win MZ, Vatalaro F, Molisch AF. Performance of low-complexity rake reception in a realistic UWB channel. In: Proc. 2002 IEEE international conference on communications, New York, NY; 2002.

[71] Win MZ, Chrisikos G, Sollenberger NR. Performance or rake reception in dense multipath channels: implications of spreading bandwidth and selection diversity order. IEEE J Sel Area Commun 2000;18(8):1516–25.

[72] Win MZ, Scholtz RA. "Characterization of ultra-wide bandwidth wireless indoor channels: A communication-theoretic view". IEEE J Sel Area Commun 2002;20(9):1613–27.

[73] IEEE P802.15-08-0780-12-0006. Channel model for body area network (BAN); 2010.

Printed in Dunstable, United Kingdom

91220871R00030